Marx Joyce
Abbott Hardy Montaigne Cooper Chesterton Emerson Austen
Defoe Melville Hugo
Machiavelli Eliot
Haggard Grimm
Carroll Christie Molière
Stoker Maupassant
Wilde Byron
Garnett Engels Schiller
Goethe Fitzgerald Hawthorne Kafka
Cotton Einstein Dostoyevsky Smith Hall
Baum Kipling Doyle Willis
Henry Nietzsche
Dumas Flaubert Turgenev Balzac
Leslie Stockton Vatsyayana Crane
Burroughs Verne
Curtis Tocqueville Gogol Vinci
Homer Widger Tolstoy Busch
Darwin Whitman
Potter Thoreau Twain Scott
Kant Freud Zola Lawrence Plato Harte
Jowett Stevenson Dickens
Andersen Burton Hesse
London Descartes
Poe Aristotle Wells Cervantes Voltaire
Hale James Hastings Cooke
Bunner Shakespeare Irving
Richter Chambers
Doré da Benedict
Chekhov Alcott
Dante Shaw Pushkin
Swift Newton
Wodehouse

# tredition®

tredition was established in 2006 by Sandra Latusseck and Soenke Schulz. Based in Hamburg, Germany, tredition offers publishing solutions to authors and publishing houses, combined with worldwide distribution of printed and digital book content. tredition is uniquely positioned to enable authors and publishing houses to create books on their own terms and without conventional manufacturing risks.

For more information please visit: www.tredition.com

## TREDITION CLASSICS

This book is part of the TREDITION CLASSICS series. The creators of this series are united by passion for literature and driven by the intention of making all public domain books available in printed format again - worldwide. Most TREDITION CLASSICS titles have been out of print and off the bookstore shelves for decades. At tredition we believe that a great book never goes out of style and that its value is eternal. Several mostly non-profit literature projects provide content to tredition. To support their good work, tredition donates a portion of the proceeds from each sold copy. As a reader of a TREDITION CLASSICS book, you support our mission to save many of the amazing works of world literature from oblivion. See all available books at www.tredition.com.

 Project Gutenberg

The content for this book has been graciously provided by Project Gutenberg. Project Gutenberg is a non-profit organization founded by Michael Hart in 1971 at the University of Illinois. The mission of Project Gutenberg is simple: To encourage the creation and distribution of eBooks. Project Gutenberg is the first and largest collection of public domain eBooks.

# Made-Over Dishes

Sarah Tyson Heston Rorer

# Imprint

This book is part of TREDITION CLASSICS

Author: Sarah Tyson Heston Rorer
Cover design: Buchgut, Berlin – Germany

Publisher: tredition GmbH, Hamburg - Germany
ISBN: 978-3-8424-6540-4

www.tredition.com
www.tredition.de

Copyright:
The content of this book is sourced from the public domain.

The intention of the TREDITION CLASSICS series is to make world literature in the public domain available in printed format. Literary enthusiasts and organizations, such as Project Gutenberg, worldwide have scanned and digitally edited the original texts. tredition has subsequently formatted and redesigned the content into a modern reading layout. Therefore, we cannot guarantee the exact reproduction of the original format of a particular historic edition. Please also note that no modifications have been made to the spelling, therefore it may differ from the orthography used today.

# CONTENTS

Preface
Stock
Cooked Fish
Meat
  Beef — Uncooked
  Beef — Cooked
  Mutton — Uncooked
  Mutton — Cooked
  Chicken — Uncooked
  Chicken — Cooked
Game
Bread
Eggs
Potatoes
  Cold Boiled
Cheese
Sauces
Salads
Cereals
Vegetables
Fruits
Sour Milk and Cream

# PREFACE

Wise forethought, which means economy, stands as the first of domestic duties. Poverty in no way affects skill in the preparation of food. The object of cooking is to draw out the proper flavor of each individual ingredient used in the preparation of a dish, and render it more easy of digestion. Admirable flavorings are given by the little leftovers of vegetables that too often find their way into the garbage bucket.

Economical marketing does not mean the purchase of inferior articles at a cheap price, but of a small quantity of the best materials found in the market; these materials to be wisely and economically used. Small quantity and no waste, just enough and not a piece too much, is a good rule to remember. In roasts and steaks, however, there will be, in spite of careful buying, bits left over, that, if economically used, may be converted into palatable, sightly and wholesome dishes for the next day's lunch or supper.

Never purchase the so-called tender meat for stews, Hamburg steaks or soups; nor should you purchase a round or shoulder steak for broiling, nor an old chicken for roasting. Select a fowl for a fricassee, a chicken for roasting, and a so-called spring chicken for broiling. Each has its own individual price and place.

Save for stock, every bone, whether beef, mutton, poultry or game, as well as all the juices that are left in the meat carving dishes on the table, and the water in which meats are boiled and in which certain vegetables are boiled. Into this storehouse—for such a stock pot is—will go also the tough ends from the rib roasts, which would become tasteless and dry if roasted; the bits that are taken from the French chops; the bone that is left on the plate from the sirloin steak; and every piece of the carcass left on the general carving plate of all sorts of game and poultry. After the meat has been taken from the roast, these bones will also be used.

# STOCK

In all good cooking there is a constant demand for a half pint or a pint of stock. Brown sauce and tomato sauce, in fact, all meat sauces, are decidedly better made from stock than water, and as it comes to every household without the additional cost of a penny, there is no excuse whatever for being without it. Save the bones collected on Saturday, Sunday and Monday. Chicken and veal bones may be kept together; beef, mutton and ham in another lot; one makes a white stock, the other brown. If the quantity is small, put them all together. Crack the bones, put them in the bottom of a large soup kettle, cover with cold water, bring slowly to boiling point and skim. Push the kettle to the back part of the stove, where the stock may simmer for at least three hours, then add an onion into which you have stuck twelve cloves, a bay leaf, a few celery tops, or a little celery seed, and a carrot cut into slices; simmer gently for another hour and strain. Tuesdays and Saturdays are the best days for making stock, as they are the days on which you have long, continuous fires; Tuesdays for ironing purposes; Saturdays for bread baking; in this way you will economize in coal, heat and time.

In making tomato soup, to each pint of tomatoes add a pint of this stock instead of water; or the stock may be simply heated, nicely seasoned and used as clear soup. By adding a little cooked rice or macaroni, you will have a rice or a macaroni soup.

In cream soups, where stock takes the place of water, less milk gives equal, perhaps better, results. For instance, in cream of celery soup, cover the celery with cold stock instead of water, using a quart instead of a pint of water, and then use only a pint of milk, having in the end the same quantity of a much more tasty soup at a less cost. One soon learns that all made-over dishes are more savory where stock is used in place of water. If peas, beans or cabbage are being cooked, this water may be added to that in which beef or mutton has been boiled, the whole reduced carefully by rapid boiling, strained and put aside for use.

# COOKED FISH

Canapés

Cold boiled fish makes excellent canapés. To each half pint of fish allow six squares of toasted bread. If you have any cold boiled potatoes left over, add milk to them, make them hot and put them into a pastry bag. Decorate the edge of the toast with these mashed potatoes, using a small star tube; put them back in the oven until light brown. Make the fish into a creamed fish. Rub the butter and flour together, add a half pint of milk, add the fish and a palatable seasoning of salt and pepper. Dish the centers on top of the toast with this creamed fish and send at once to the table. A very little fish here makes a good showing, and is one of the nicest of the hot canapés.

Baked Sardines

After sardines have once been opened it is best to remove them from the can and make them into some dish for the next meal. They may be broiled and served on toast, or made with bread crumbs into sardine balls and fried, or baked. To bake them, stir the oil from the can into a half cupful of water, add a teaspoonful of Worcestershire sauce, a half teaspoonful of salt and a dash of pepper. Put the fish into a baking pan, run them into the oven until very hot, then dish them, baste them with the sauce and send them at once to the table.

Fish Croquettes

Any cold boiled fish that is left over may be made into croquettes. To each cupful of the cold fish allow one level tablespoonful of butter, two level tablespoonfuls of flour and a half cupful of milk. Rub the butter and flour together, add the milk; when boiling take from the fire. Add to the fish a level teaspoonful of salt, a dash of black pepper, a tablespoonful of chopped parsley and a few drops of onion juice; mix this carefully with the paste and turn out to cool. When cold, form into small cylinders, dip in beaten egg and fry in deep hot fat.

Fish à la Crême

One pint of cold boiled fish, mixed with a half pint of white sauce. Turn this into a baking dish and brown. Or when the two are carefully heated together, serve in either ramekin dishes or in a border of browned mashed potatoes.

# MEAT

As meat is the most costly and extravagant of all articles of food, it behooves the housewife to save all left-overs and work them over into other dishes. The so-called inferior pieces—not inferior because they contain less nourishment, but inferior because the demand for such meat is less—should be used for all dishes that are chopped before cooking, as Hamburg steaks, curry balls, kibbee, or for stews, ragouts, pot roasts and various dishes where a sauce is used to hide the inferiority and ugliness of the dish. We have no occasion here to spend money on good looks.

If one purchases meat for soup, the leg and shin are the better parts. This, however, is not necessary in the ordinary family, as there are always sufficient bones left over for daily stock. All meat left over from beef tea, tasteless as it is, may be nicely seasoned and made into curries or into pressed meat, giving again a nice dish for lunch or supper. Remember, that where the flavoring of the beef has been drawn out into the water, as in making beef tea, another decided flavor must be added to make the made-over dish palatable. For this reason, curries, pressed meats, served with either Worcestershire or tomato sauce, are chosen.

Cold mutton may be made into pilau, hashed on toast with tomato sauce, hashed with caper sauce, made into escalloped mutton, barbecued mutton, casserole, or macaroni timbale; all sightly dishes, quite handsome enough to place before the choicest guest. Spiced meats, as beef *à la mode*, may be served cold with cream horseradish sauce and aspic jelly. If warm, they will be made into ragouts, or some form of dish with a brown or tomato sauce. It is well to bear in mind that white meats will be served with white or yellow sauces; dark meats with brown or tomato sauces. The coarse tops of the sirloin steak, the tough end of the rump steak, if broiled, cannot possibly be eaten, as the dry heat renders them difficult of mastication. Cut them off before the steak is broiled, and put them aside to use for Hamburg steaks, curry balls, timbale or cannelon, making a new and sightly dish from that which would otherwise have been thrown away.

If you use ham, and have had a piece boiled, after the even slices are taken off, chip the remaining tender pieces for frizzled ham, making it as frizzled beef is made. The bits around the bone that cannot possibly be sliced, will be chopped and made into potted or deviled ham. Throw the bone into the stock pot.

A meat chopper or grinder, which costs but a dollar and a half or two dollars, will save its price in the utility of these scraps in less than a month.

The water in which you boil a leg of mutton, chicken, turkey or a fresh beef's tongue, or such vegetables as string beans, peas, rice, macaroni or barley, put aside and use in place of plain water to cover the bones for stock-making. The water in which cabbage is boiled should be saved alone and used the next day for a soup Crécy; the flavor of the cabbage, with a carrot that has been slightly browned in butter, makes a delightful soup without the addition of meat.

# BEEF—UNCOOKED

The uncooked tough bits or pieces of beef may be made into any of the following dishes:

Kibbee

Chop uncooked tough meat very fine; put it twice through a grinder. To each pound, allow a tablespoonful of grated onion, a tablespoonful of chopped parsley, a teaspoonful of salt, just a dash of pepper, and a half cup of toasted piñon nuts. Form into balls about the size of an egg, stand in a baking pan, add a half pint of strained tomatoes, a tablespoonful of butter, and bake slowly thirty minutes, basting three or four times. If more than one pound of meat is used, all the ingredients must be increased accordingly.

Hamburg Steaks

The genuine Hamburg steaks are rich in onion and very rich in fatty matter, too much so to be wholesome; so we will modify them, that they may be eaten even by dyspeptics or persons with weak digestion. Put twice through a meat chopper the tough ends of steaks or bits of the round. To each pound of this meat allow a half teaspoonful of celery seed, a teaspoonful of grated onion. Form into thick even cakes, being sure that the center and sides are the same thickness. These may now be broiled over a clear fire, or under the gas lights in your gas broiler, or they may be dropped into a thoroughly heated iron pan. As soon as browned on one side, turn and brown the other. If the steaks are an inch thick, it will take eight minutes for perfect cooking. An exceedingly satisfactory way is to brown them quickly over a hot fire, then put the pan in the oven and allow them to cook for five minutes. Dust with salt, season with a little butter and pepper, and send to the table on a very hot dish; or serve with brown or tomato sauce. If they have been cooked over the fire, or in the oven, put a tablespoonful of butter into the pan in which they were cooked, add a tablespoonful of flour, a half cup of stock, and a half cup of strained tomatoes. When boiling, add a teaspoonful of salt, a dash of pepper, and pour over the steaks.

Cannelon

Put twice through the meat chopper one pound of tough meat, season with a teaspoonful of salt, a dash of pepper, and, if you like, a little celery seed or chopped celery top; take this chopped meat into your hands, and form it into a roll about four inches in diameter and six inches long. Roll this in a piece of oiled paper, put it in a baking pan, bake in a quick oven thirty minutes, basting the paper with melted butter three or four times. When done, remove the paper, dish the cannelon, and pour around plain tomato sauce.

Brown stew

Cut any left-over pieces of uncooked tough meat into cubes of one inch. Put a couple of tablespoonfuls of suet into a saucepan; when rendered out, remove the cracklings. Dust the bits of meat with a tablespoonful of flour, throw them into the hot suet, and shake until brown. Draw the meat to one side, and add to the fat in the pan a second tablespoonful of flour; mix, add one pint of water or stock, stir until boiling, add a teaspoonful of salt, a bay leaf, slice of onion, a teaspoonful of browning or kitchen bouquet; cover and simmer gently until the meat is tender, about an hour and a half. The proportions given here are for one pound of beef. This may be served plain, or in a border of rice, or with dumplings. If dumplings, put a pint of flour into a bowl, add a teaspoonful of salt and one of baking powder; mix thoroughly and add sufficient milk to just moisten; drop by spoonfuls over the top of the stew, cover the saucepan and cook for ten minutes. Do not lift cover during the ten minutes or the dumplings will fall.

Beef Timbale

Chop fine any left-over tough bits of lean beef. Cook together for a moment a gill of strained tomatoes and one cup of bread crumbs; add to the meat, rub to a smooth paste, season with a quarter of a teaspoonful of celery seed, a half teaspoonful of salt and a dash of pepper; mix, and then stir in carefully the well-beaten whites of two eggs; fill into custard cups, stand in a pan of boiling water, and cook in a moderate oven twenty minutes. Serve with tomato sauce. This recipe is for one pound of beef.

# BEEF—COOKED

Ragout

Cut pieces of cold boiled or roasted beef into cubes of one inch; to each quart of this allow two tablespoonfuls of butter, two of flour and a pint of stock. Rub the butter and flour together, add the stock, stir until boiling; add a tablespoonful of onion juice, a teaspoonful of browning or kitchen bouquet, a teaspoonful of salt, a tablespoonful of tomato catsup, a tablespoonful of chopped parsley; add the meat; stand over the back part of the stove until thoroughly hot; serve on a heated platter garnished with triangular pieces of toasted bread. A few left-over olives, mushrooms, or even a chopped truffle, may be added.

Bresleau

Chop sufficient cold cooked meat to make one pint, season it with a teaspoonful of salt and a quarter of a teaspoonful of pepper. Put a half cup of stock or water, two tablespoonfuls of bread crumbs and a tablespoonful of butter over the fire; when hot, add to it the meat; take from the fire and stir in carefully two well-beaten eggs. Put this in greased custard cups, stand them in a baking pan half filled with boiling water, and bake in a moderate oven fifteen or twenty minutes; serve with tomato sauce or sauce Béchamel.

Beef Croquettes

Chop sufficient cold cooked beef to make one pint; add to it a teaspoonful of salt, a teaspoonful of onion juice, a dash of cayenne, a quarter of a teaspoonful of pepper, and a grating of nutmeg. Put a half pint of milk over the fire. Rub together one tablespoonful of butter and two tablespoonfuls of flour, add them to the hot milk, stir until you have a smooth thick paste; take from the fire; mix with it the meat, and turn out to cool. When cold, form into croquettes.

Beat one egg, add to it a tablespoonful of warm water, and beat again. Dip the croquettes first into this, then roll them in bread crumbs, and fry them in smoking hot fat. They may be served plain or with tomato sauce.

### Beef Steak Pudding

Cut cold cooked steak into cubes of a half inch. To each pint of these allow a half pint of milk, six tablespoonfuls of flour, two eggs, and two tablespoonfuls of chopped suet. Put the flour into a bowl; beat the eggs, add to them the milk, then add gradually to the flour; make perfectly smooth. Cover the bottom of a baking dish with a layer of the batter, put in the bits of steak, sprinkle over the chopped suet, then a dusting of salt and pepper, and, if you like, a few drops of onion juice; now put over the remaining quantity of the batter, and bake in a moderately quick oven an hour and a half.

### Potato Dumplings

Take any pieces of cold cooked meat, chop them fine, season carefully with salt, pepper, chopped parsley or celery. To each pint allow two tablespoonfuls of melted butter. For the crust you may use left-over cold mashed potatoes; if so, add a little milk and stir them over the fire until smooth and hot. If potatoes are boiled for the purpose, add salt, butter and milk, and beat them until light. Line to the depth of one inch, a baking dish, put the meat in the center, cover the top with mashed potatoes, smooth, brush with milk and bake in a moderate oven a half hour.

### Gobbits

Scrape and cut into fancy pieces one good-sized carrot and one turnip. Put these into a saucepan, cover with a pint of stock, and cook slowly until the vegetables are tender. Have ready, cut into cubes of one inch, sufficient cold cooked beef to make a quart; add it to the vegetables, simmer a few minutes until the meat is hot; have ready also one cup of rice that has been boiled thirty minutes in clear water, drained and dried. Arrange this in a border around the meat dish. Put two tablespoonfuls of butter and flour into a saucepan; mix. Drain the liquor from the meat and vegetables, which should now measure one pint; if not, add sufficient stock to make a pint; add this to the butter and flour, and stir until boiling. Dish the meat and vegetables in the centre of the rice border. Take the sauce from the fire, add a teaspoonful of salt, a dash of pepper and the yolks of two eggs. Reheat for just an instant, strain over the meat mixture, dust with chopped parsley, and serve at once.

Beef Fritters

Chop sufficient cold cooked beef to make one pint; add to it a teaspoonful of salt, and a quarter of a teaspoonful of pepper. Beat two eggs until light, add to them a half pint of water or stock; stir into this one and a half cups of flour, beat until smooth, add a teaspoonful of baking powder and the meat. Drop this by spoonfuls into smoking hot fat; cook about three minutes, drain on brown paper, and serve either on a folded napkin, or in a dish with tomato sauce.

Minced Beef on Toast

Take the meat from between the bones of a rib roast, or any little bits that would not be serviceable in other dishes, chop them fine, and to each pint, allow one tablespoonful of butter, one of flour and a half pint of tomatoes or stock. Mix the butter and flour together, then add the tomatoes strained or stock; when boiling add the meat, and a palatable seasoning of salt and pepper. Stand the mixture over hot water until smoking hot, and serve on squares of toasted bread.

Barbecue of Cold Beef

Cut cold-roasted or boiled beef into thin slices. Put into your saucepan two tablespoonfuls of butter, two tablespoonfuls of catsup and two tablespoonfuls of sherry; stir until hot; drop the slices of beef into this, cover the saucepan, shake occasionally for a minute, until the beef is smoking hot, and send at once to the table. This is exceedingly nice made and served from a chafing dish. This dish may be made by omitting the sherry and using a teaspoonful of Worcestershire sauce, a teaspoonful of mushroom catsup and two tablespoonfuls of stock.

Salt Beef Hash No. 1

Cold cooked corned beef is best made into hash. Chop sufficient to make one pint. Chop the same quantity of cold boiled potatoes; mix the two together, put them into a saucepan, add a half pint of stock, a tablespoonful of butter, teaspoonful of onion juice and a quarter of a teaspoonful of black or white pepper. Stir carefully and constantly until the mixture reaches the boiling point. Serve at once on buttered toast.

Salt Beef Hash No. 2

Chop enough cold cooked corned beef to make a pint; chop the same quantity of cold boiled potatoes; mix the two together. Put them into a stewing pan, add one pint of stock; simmer for just a moment; take from the fire, add two eggs well beaten, a dash of pepper; turn the mixture into a baking dish and bake in a quick oven twenty minutes.

Rechauffee of Beef

Cut any left-over cold beef into thin slices. Cut into slices three cold boiled potatoes. Peel two tomatoes, cut them into halves, squeeze out the seeds, and then cut the tomatoes into small bits. Chop one good sized onion. Put a layer of tomato in the bottom of a baking dish, then beef, then a seasoning of onion, salt and pepper, and if you have it, a little chopped celery, then potatoes, then again tomatoes, beef, and so continue until you have used the materials, having the last layer tomatoes. Dust the top with bread crumbs, put over a few bits of butter and bake a half hour in a moderately quick oven.

Steak Pudding

Cut any cold left-over steak into thin slices, and cut these slices into bits one inch long. Put one quart of flour in a bowl, and add to it one cupful of chopped uncooked suet. Chop the suet and flour together for a minute, add a level teaspoonful of salt, a saltspoonful of black pepper, and sufficient cold water to just moisten. Take the dough on the board and roll it out into a sheet; make it a little larger than an ordinary pie dish. Season the bits of meat, put them on one-half the sheet, lay over the top twelve good fat oysters, brush the under half of the dough with the white of egg or water; fold over the other half and make two or three holes in the top. Put it in a cheese cloth and steam for two hours. Remove the cloth, brush the pudding with the yolk of the egg and bake in a quick oven a half hour.


Panada of Beef

Chop sufficient cold cooked beef to make one pint; season it with a teaspoonful of salt, a tablespoonful of chopped parsley and a dash of pepper. Put this in the bottom of a baking dish. Crush six Uneeda biscuits, pour over them a half pint of milk, let them stand a minute or two, add one egg, well beaten, a half teaspoonful of salt and a saltspoonful of pepper. Pour this over the beef and bake in a moderate oven twenty minutes to a half hour.

Other meats may be substituted for beef.

# MUTTON — UNCOOKED

Tough pieces of uncooked mutton may be put twice through the meat chopper and used for curry balls or for stuffing for tomatoes or egg plant; in fact, in almost any way that one would serve uncooked beef. Having fewer pieces of uncooked scrap mutton than of beef, we are less accustomed to seeing them used.

Curry Balls

Put any pieces of tough uncooked mutton twice through the meat chopper; season the meat with salt, pepper and onion juice. Form into little balls the size of an English walnut. Put two tablespoonfuls of butter into a saucepan; when hot, throw the balls into the butter, and shake until carefully browned. Lift them from the saucepan, and to the butter in the pan add a teaspoonful of curry, a tablespoonful of flour, mix and add a half pint of stock; stir carefully until boiling; pour this over the balls, cook, slowly for twenty minutes, add two tablespoonfuls of lemon juice and serve in a border of rice. Cocoanut milk may be used instead of stock.

# MUTTON — COOKED

While mutton belongs to the red meats, when carefully cooked it may be used in many ways in which you would use chicken or veal. Capers and tomato, with a slight flavoring of mint, are more agreeable with mutton than with almost any other meats.

### Bobotee

Chop sufficient cold boiled mutton to make a pint. Put two tablespoonfuls of butter and one onion sliced into a saucepan; stir until the onion is slightly brown; then add a half pint of stock or milk and four tablespoonfuls of bread crumbs. Stand this on the back of the stove for about five minutes while you blanch and chop fine a dozen almonds. Add these to the meat, then add a teaspoonful of curry powder, and a teaspoonful of salt. Beat three eggs until light, stir them into the meat, then turn the whole into the saucepan. Rub the bottom of the baking dish first with a clove of garlic, then sprinkle over a tablespoonful of lemon juice and put here and there a few bits of butter; put on this the mixture, and bake in a quick oven twenty minutes. Serve in the dish in which it is baked, and pass with it plain boiled rice.

### Boudins

Chop sufficient cold cooked mutton to make a pint. Put a half cup of stock, two tablespoonfuls of bread crumbs and a tablespoonful of butter over the fire. When hot, take from the fire, add the meat and three eggs well beaten; add a teaspoonful of salt and a dash of pepper. Put the mixture into greased custard cups, stand in a baking pan half filled with boiling water, and cook in a moderate oven fifteen to twenty minutes. Serve with sauce Béchamel. The bottom of the cups may be garnished with chopped mushrooms, capers, or chopped truffles, or dusted thickly with chopped parsley.

### Klopps

Chop sufficient cold boiled mutton to make a pint; add to it a half pint of bread crumbs and sufficient white of egg to bind the whole together; add a teaspoonful of salt and a dash of white pepper. Form into balls the size of English walnuts; drop into a kettle of boiling water; pull the kettle to one side of the fire where it cannot possibly boil, and cook the klopps slowly for five or six minutes. When done they will float on the surface. Lift, drain carefully, put on to a heated dish, pour over cream celery or cream oyster sauce, and serve with them peas and boiled rice.

### Curry of Mutton

Put two tablespoonfuls of butter and one sliced onion into a pan; cook slowly until the onion is perfectly tender; add one clove of garlic mashed, a teaspoonful of curry powder and a teaspoonful of turmeric; mix thoroughly, add a half pint of stock, or, better, cocoanut milk; stir until boiling, add one quart of cold cooked mutton chopped fine; heat thoroughly, add a tablespoonful of lemon juice, and pour at once into a platter that has been garnished with boiled rice.

### Mutton with Anchovy

Chop sufficient cold boiled mutton to make one pint; mash fine three anchovies. Put two tablespoonfuls of butter into a saucepan, add one sliced onion, cook until the onion is soft and yellow, add a clove of garlic mashed, add to this the anchovies and a half pint of stock; simmer gently for fifteen minutes, and press through a sieve. Add a tablespoonful of capers, two or three leaves of mint that have been bruised, and the mutton chopped fine. Heat over boiling water for fifteen minutes, and serve on squares of toasted bread. This may be served plain or the top of each piece may be capped with a carefully poached egg.

### Pilau

Cut into bits any pieces of cold cooked mutton; put them into a saucepan, cover with water, add a grated onion, a bay leaf and two or three cardamom seeds. Sprinkle over a half cup of rice that has been carefully washed; cover the kettle and simmer slowly until the rice is tender. Dish the mutton, putting the rice over the top, cover the whole with a nicely made tomato sauce, and send at once to the table.

Mutton Salad

Any pieces of cold-roasted or boiled mutton may be cut into dice and used for an ordinary mutton salad. At serving time arrange this neatly on lettuce leaves, or any accessible green; season with salt and pepper, and cover with mayonnaise dressing to which has been added a tablespoonful of capers.

Where celery, lettuce or other fresh greens cannot be procured, canned asparagus may be mixed with the mutton or may be served with it as a garnish; giving an exceedingly agreeable accompaniment. Where asparagus cannot be obtained, a can of peas may be drained, washed, drained again, and added to the mutton before it is mixed with the mayonnaise dressing, or the mutton may be mixed with mayonnaise and filled into tomatoes that have been peeled and the centers scooped out. Stand each on a little nest of lettuce leaves or on a bunch of cress, and garnish the top with capers.

French Lamb Stew

1 quart of bits of cold left-over lamb or mutton 1 pint of green peas 1 quart of water 3 stalks of mint 1 teaspoonful of onion juice 1 teaspoonful of salt 1 saltspoonful of pepper

Put the lamb, water and all the seasoning into a saucepan. Shell and wash the peas, put them over the top, cover the pan and bring quickly to a boil, lift the lid, and boil rapidly twenty minutes until the peas are tender. Rub together the butter and flour, stir them carefully into the stew, bring again to boiling point and serve.

Lamb Stew with Tomatoes

Follow the preceding recipe, using a quart of strained tomatoes in place of a quart of water.

# CHICKEN – UNCOOKED

In purchasing a chicken for timbale, select a large one, but not an old fowl. After the chicken has been drawn, remove the white meat, which is used uncooked for timbales. The dark meat may be cooked at once and utilized for boudins, croquettes, salad, cecils, creamed hash, or served on toast with sauce Bordelaise, or used in chafing dish next day. Or if you prefer to use it raw, devil the legs and use the bones for soup.

Timbale

Chop fine the uncooked white meat of a chicken; this should weigh a half pound. Then rub it with the back of a wooden spoon against the side of a bowl until perfectly smooth. Put one cup of white bread crumbs and a half cup of milk over the fire; stir until boiling; when cold, rub this thoroughly with the meat, and press it through an ordinary flour sieve. Stir into it carefully the well-beaten whites of five eggs, add a teaspoonful of salt, a dash of white pepper; fill into greased timbale cups, stand in a baking pan of boiling water, cover with oiled paper, and bake in a moderate oven fifteen to twenty minutes. The bottoms of the cups may be garnished with chopped truffle, chopped mushrooms, chopped parsley, or nicely cooked green peas. Serve with the timbales either a plain cream sauce or a cream mushroom sauce. Peas are the usual accompaniment.

Or the timbale molds may be lined with this mixture, and the centers filled with creamed mushrooms; put enough of the timbale mixture over the top to hold in the stuffing; they will then be cooked and served in the usual manner.

Deviled Chicken Legs

Carefully remove the bones from the legs of an uncooked chicken. To a half cup of bread crumbs add twelve chopped almonds, two tablespoonfuls of toasted piñon nuts, a tablespoonful of parsley, a half teaspoonful of salt and a dash of cayenne; moisten with two tablespoonfuls of butter. Stuff this into the spaces from which you have taken the bones, tie the legs top and bottom to keep in the stuffing. Place the bones from the carcass of the chicken in the soup kettle, cover with cold water, and when the water reaches boiling point place the legs on top of the bones and cook continuously for two hours. They may be served hot with sauce, or cold, cut into thin slices garnished with aspic.

English Chicken Balls

Chop fine the dark meat left over from timbales, add a half can of finely chopped mushrooms, a teaspoonful of salt, a half teaspoonful of pepper, a tablespoonful of chopped parsley, a dozen blanched and finely chopped almonds and one raw egg; mix thoroughly and form into balls the size of an English walnut. Arrange these over the bottom of a saucepan, cover with stock, add a bay leaf, a slice of onion and of carrot; cook slowly a half to three-quarters of an hour; drain, saving the stock. Dish the balls in the center of a platter, put around the edge a row of potato bullets, outside of that small triangles of toast. Put a tablespoonful of butter and one of flour into a saucepan; mix, add a half pint of stock in which the balls were cooked, stir until boiling, take from the fire, add the yolk of one egg beaten with two tablespoonfuls of cream; add a half teaspoonful of salt and a dash of pepper; strain this over the balls and serve.

# CHICKEN — COOKED

The remains of cold chicken or turkey may be used in precisely the same manner, or made into croquettes, using the same rule as for beef croquettes. With an accompaniment of mayonnaise of celery, or mayonnaise of tomato, they make an extremely good luncheon dish. For an evening entertainment they may be simply garnished with cooked peas. Meat croquettes are usually made into pyramid forms; they may, however, be made into cylinders. Boudins of chicken or turkey are also exceedingly nice.

Creamed Hash on Toast

This is one of the tastiest of all the warmed-over chicken dishes. Chop the chicken fine, and to each pint allow one tablespoonful of butter, one of flour and a half pint of milk. Rub the butter and flour together, add the milk, stir over the fire until boiling, season the meat with a teaspoonful of salt and a dash of pepper, add to the milk sauce, and stir over hot water for fifteen minutes. The flavoring may be changed by adding three or four chopped mushrooms, or, if you have it, a chopped truffle; but it is exceedingly good plain. Heap this on squares of nicely toasted bread, serve at once, or you may garnish the tops with carefully poached eggs.

Casserole

Wash a half cup of rice; throw it into boiling water, boil for twenty minutes, drain, add a half cup of milk, a tablespoonful of butter, a level teaspoonful of salt and a quarter of a teaspoonful of pepper; stir until you have a rather smooth thick paste. Brush custard cups, line them to the depth of a half inch with this rice mixture; make a plain milk sauce, as in preceding recipe, and add a pint of seasoned chicken. Fill the space in the rice cups with this cream mixture, put over a covering of rice, stand the cups in a pan of boiling water, and bake in a moderate oven for twenty to twenty-five minutes. Turn these carefully on a heated dish, pour around cream sauce and serve. They may be garnished with green peas, mushrooms or truffles. While this is an exceedingly economical dish it is at the same time an elegant one.

### Indian Hash

Chop fine sufficient cold-roasted duck, chicken, or turkey to make one pint. Cut a good-sized onion into very thin slices. Pare, core, and chop fine one apple. Put two tablespoonfuls of butter in a saucepan, add the apple and the onion; toss until brown, then add not more than an eighth of a teaspoonful of powdered mace, a half teaspoonful of salt, a teaspoonful of curry powder, a tablespoonful of flour, a teaspoonful of sugar; mix and add a half pint of stock or water; now add the meat, stir constantly until smoking hot, then stand over hot water, covering closely for twenty minutes. Add two tablespoonfuls of lemon juice and serve in a border of rice.

### Mock Terrapin or à la Newburg

Pieces of cold-roasted chicken, turkey or duck may be used for making terrapin or à la Newburg. Cut the meat into pieces of fairly good size; measure, and to each pint of this allow a half pint of sauce; rub together two tablespoonfuls of butter and one of flour. Rub to a smooth paste the hard boiled yolks of three eggs; add to the butter and flour a gill and a half (three-quarters of a cup) of milk; stir until smoking hot. Do not let the mixture boil; then add this a little at a time to the yolks of the eggs, rubbing until you have a perfectly smooth golden sauce; press this through a sieve. Before beginning the sauce, sprinkle the chicken with four tablespoonfuls of sherry or Madeira, the latter preferable. Add the chicken to the sauce, stir until each piece is thoroughly covered; add a half teaspoonful of salt, just a drop of extract of nutmeg or a grating of nutmeg, an eighth of a spoon of white pepper (black pepper, of course, may be used); cover and stand over hot water, stirring occasionally until the mixture is smoking hot.

Chicken Supréme

This may be made from either chicken or turkey cut into dice; add an equal quantity of canned mushrooms; for instance, to one pint of cold chicken, add one can of mushrooms. Put two tablespoonfuls of butter and two of flour in a saucepan; mix without browning, then add two cups (one pint) of chicken stock; stir constantly until boiling, add two tablespoonfuls of thick cream, and the yolks of four eggs; strain, add the chicken and mushrooms, a level teaspoonful of salt, a quarter of a teaspoonful of white pepper, ten drops of celery extract or just a little celery seed. Stand this mixture over hot water, watching carefully until it is thoroughly heated; remember that any boiling will curdle the egg. Serve this on a heated dish either in a border of rice or garnished with squares of toasted bread. This mixture is also served in bread patês, or it may be served in chicken muffin cases.


Chicken Cutlets

Chop cold cooked chicken or turkey very fine; to each pint allow a half can of mushrooms chopped fine. Put one tablespoonful of butter and two of flour into a saucepan, mix, and add a half pint of chicken stock. When smooth and thick take from the fire, add the yolks of two eggs, the chicken and mushrooms, a teaspoonful of salt, quarter of a teaspoonful of pepper, a teaspoonful of onion juice, a grating of nutmeg and a tablespoonful of chopped parsley; stir over the fire for a moment; turn out to cool; when cold form into cutlet-shaped croquettes, dip in egg and bread crumbs, and fry in smoking hot fat. These may be served plain, with a garnish of peas, or they may be served with sauce Béchamel.


Duck Bordelaise

Portions of cold duck may be cut into convenient pieces, sprinkled with wine, about four tablespoonfuls to the pint, and allowed to stand while you make sauce Bordelaise. Put one tablespoonful of butter and one of flour into a saucepan; mix, add a teaspoonful of browning or kitchen bouquet and a half pint of stock; stir until boiling, add a tablespoonful of grated onion, a half teaspoonful of salt, a dash of pepper and, if you have it, a tablespoonful of finely-chopped ham; cook for five minutes and strain; add three or four fresh mushrooms or a half dozen canned mushrooms and the duck. Stand over boiling water until the mixture is thoroughly heated. Send to the table garnished with triangles of toasted bread. A few stoned olives or sliced olives may be added in the place of the mushrooms, and you would then have salmi of duck.

# GAME

Bits of cold broiled or roasted game may be chopped very fine, rubbed to a smooth paste either in a bowl or mortar. To each half pint of this mixture allow two tablespoonfuls of brown sauce thoroughly rubbed with the game, and the unbeaten white of one egg; press the whole mixture through an ordinary flour sieve; then stir in the well-beaten whites of two eggs, four mushrooms chopped almost to a powder, and a seasoning of salt and pepper. Fill this into little greased molds or cups; the cups may be garnished with chopped truffle or mushrooms, or served plain. Fill in the mixture, stand the cups in a baking pan half filled with boiling water; cook in a moderate oven twenty minutes. The little bomb-shaped molds are the better sort to use for these. Serve with brown sauce either plain or flavored with mushrooms.

# BREAD

The better way is to cut just sufficient bread for each meal so that there will be really no left-overs. If, however, a few slices are accidentally left, put them aside in a can or jar, never in the regular bread box with the bread; one or two slices will invariably be missed until sufficiently old to mold and contaminate the remaining quantity of bread in the box, and then, too, they are more apt to accumulate in this way than in a separate box. The neater pieces may be used for toast for breakfast or lunch or supper. The next best pieces use for bread and butter custard; the crusts dry, roll and put aside to be ready for breading articles to be fried, or for escalloped dishes. In this way every piece, no matter what its condition, will be utilized.

Bread and Butter Custard

Beat two eggs, without separating, until light, add four tablespoonfuls of sugar and a pint of milk, mix and add a grating of nutmeg; turn into an ordinary baking dish, cover the top with buttered bread, butter side up; bake in a moderate oven just as you would a cup custard, until you can put a spoon handle down in the center of the custard and it will come out free from milk.

Little Puddings à la Grand Belle

Roll slices of stale bread into fine crumbs. Brush small custard cups, or a border mold with melted butter, sprinkle over a few currants or raisins, or any fruit that you may have left over. Fill the cups with crumbs. Beat three eggs, without separating, until light; add three tablespoonfuls of sugar, a teaspoonful of vanilla and a pint of milk. Pour this carefully over the bread crumbs, let them stand for about five minutes until the mixture has been soaked up and the bread crumbs soft; then stand in a pan of boiling water, cover with oiled paper and cook in the oven a half hour. Turn out and serve hot with egg sauce.

### Bread Croquettes

Rub sufficient stale bread to make one quart of crumbs; add four tablespoonfuls of sugar, a half cup of cleaned currants, or any fruit that you have left over, and a grating of nutmeg; sprinkle over a teaspoonful of vanilla, and add sufficient beaten eggs (about three) to moisten the crumbs. Form into small cylinder-shaped croquettes, dip in egg and roll in bread crumbs and fry in smoking hot fat. Serve hot with sugar sauce.

### Bread Muffins

Cover a quart of bits of bread that have been broken apart, with one pint of milk; soak for fifteen minutes, then with a spoon beat until you have a smooth paste; add the yolks of three eggs, a tablespoonful of melted butter and one cup of flour that has been sifted with a heaping teaspoonful of baking powder. Fold in carefully the well-beaten whites of the eggs, and bake in muffin pans in a quick oven about twenty minutes.

Muffins left from breakfast may be pulled apart and toasted for lunch or supper. Pieces of stale sponge cake, in fact, any stale cake may be used for cabinet puddings, for cream puddings, or for croquettes.

# EGGS

The soft boiled eggs that are left from breakfast will be at once hard boiled, put into the refrigerator, and when four have accumulated, use them for Beauregard eggs, à la Newburg dishes or garnishes. Poached eggs that are left over may be dropped at once into boiling water, cooked slowly until perfectly hard, and put aside for chopping, to use as a garnish for a curry or some vegetable dish with which they will nicely blend.

The tablespoonful or two of stewed tomatoes left in the dish from dinner will be put aside to use for tomato omelet, or they may be added to the roasted beef gravy for dinner, converting a plain homely gravy into one of better flavor. The half cup of peas may be added to to-morrow's consommé, or used as a garnish for the breakfast omelet. The green portions of celery will be put aside for stewing; the tender white part for serving raw; while the leaves and roots will be used for flavoring soups and sauces.

The yolk of egg left over, if put into a cup or saucer will, in less than two hours, become hard, dry and useless. This same yolk dropped into a cup half filled with cold water will keep for several days, and may be used for mayonnaise or added to a sauce. When needed, it may be carefully lifted with a spoon and used the same as a fresh yolk.

### Whites of Eggs

The yolks of eggs are quite easily disposed of, as sauces frequently call for the yolk of one or two eggs; then they may be used for mayonnaise dressing, or added to various dishes. The whites of eggs, however, accumulate. One of the ways of getting hard-boiled yolks, without wasting the whites, is to separate the white and the yolk before the egg is cooked; drop the yolk down into a kettle of boiling water; then stand on the back part of the stove for fifteen or twenty minutes until it is hard. The yolk will cook in this way just as well as with the white in the shell. Now, you have the uncooked whites, which may be used for a simple white cake, apple float, soufflés, plain or with fruit.

### Beauregard Eggs

Separate the whites and yolks of five hard-boiled eggs, press through an ordinary fruit press, or chop very fine. Make a half pint of cream sauce; when boiling, add the whites of the eggs. Have ready on a heated platter five squares of toasted bread; heap the white sauce over these squares, dust the top with the yolks of the eggs, then with a little salt and pepper, and send at once to the table.

### Egg Croquettes

Put five hard-boiled eggs through a vegetable press, or chopper. Put one tablespoonful of butter and two of flour into a saucepan, add a half pint of milk, stir until boiling, add a half cup of stale, unbrowned bread crumbs, a teaspoonful of salt, a tablespoonful of chopped parsley, a dash of pepper and a half teaspoonful of onion juice; add the eggs, mix and turn out to cool. When cold form into cutlets, dip in egg and then in bread crumbs and fry in smoking hot fat. Serve with plain cream sauce. These with peas make an exceedingly nice luncheon dish.

### Gold Cake

One frequently has four or five yolks left after having used the whites for some light dish, as mock charlotte. Beat a half cupful of butter to a cream, add gradually one cupful of sugar. When very, very light, add the yolks of the eggs and beat for ten or fifteen minutes; then add one cupful of water, and two and a half cupfuls of flour, sifted with three level teaspoonfuls of baking powder. Beat thoroughly, and bake in a small round or square pan.

### German Slaw

This will use the yolks of two eggs and any little sour cream that may be left over. Shred the cabbage and soak it in cold water, changing the water once or twice. When crisp, wring it perfectly dry in a towel. Beat the yolks of two eggs, add a half cupful of sour cream, four tablespoonfuls of vinegar; stir this over the fire until it thickens. Take from the fire, add a half teaspoonful of salt and a dash of pepper; mix it with the cabbage and turn it into the serving dish. This quantity of dressing will be quite sufficient for about one quart of cabbage.

Apple Snow

In making sauce Hollandaise or mayonnaise one always has quite a quantity of the left-over whites. These may be made into various sponges, or used for fruit snow. Beat the whites of four or five eggs until light, then add two level tablespoonfuls of sifted powdered sugar to the white of each egg and beat until dry and glossy. Grate into this one tart apple, fold it quickly, float it on a little dish of good milk or cream, and send it at once to the table. If you have one or two little stale cakes, or a bit of sponge cake, stale, grate it, dust the top, and if you have just a little jelly, you may dot it here and there with the jelly. This must be made just before the dinner hour, or the apple will lose its color. Grated pear, or two or three peaches pressed through a sieve, or one or two soft bananas may be beaten and used in the place of the apple.

# POTATOES

Cold baked potatoes will be converted at once into stuffed potatoes, and put aside for rewarming. Two cold boiled potatoes will make a comfortable dish of hashed browned potatoes, or may be served with cream sauce or au gratin.

Stuffed Potatoes

Baked potatoes that are left over must be made into stuffed potatoes before they are heavy and cold. At the close of the meal at which they were first served, cut the potatoes directly into halves, scoop out the inside portion, put it through an ordinary vegetable press, or mash it fine; add a little butter, salt, pepper and sufficient milk to make a light mixture; stand this over hot water and beat until light and smooth. Put it back into the shells, and stand them aside in a cold place. When ready to serve, brush the top with beaten egg, run them into a quick oven until hot and golden brown.

Potato Croquettes

Cold mashed potatoes may be made into croquettes by adding to each pint four tablespoonfuls of heated milk, the yolks of two eggs, a tablespoonful of chopped parsley, a teaspoonful of grated onion, a quarter of a teaspoonful of pepper; stir over the fire until the mixture is thoroughly heated; form into cylinder-shaped croquettes, dip in egg and rolled bread crumbs and fry in smoking hot fat. Potato croquettes are more difficult to fry than meat croquettes; the fat must be at least 365 degrees (Fahr.) and the rolling carefully done.

Potato Puff

The above mixture may have the whites of the eggs beaten and stirred in, and baked in the oven; serve in the same dish in which it was baked.

### Potato Roses for Garnishing

Cold boiled potatoes may have added sufficient milk to make a soft paste; stir it over the fire until smooth; put it into your pastry bag, using a star tube; hold the bag firmly, pressing out on greased papers these little potato roses; brown in the oven and use them for garnishing fish dishes.

### Potato Custards

Stir two cups of cold mashed potatoes, with four tablespoonfuls of milk, over the fire until they are warm and light; take from the fire and add three eggs beaten light with four tablespoonfuls of sugar. Add a teaspoonful of vanilla, stir in carefully a pint and a half of milk. Put this mixture into greased custard cups; stand in a baking pan of boiling water and bake in a moderate oven until set, about twenty or thirty minutes.

Where a little cooked meat and, at the same time, mashed potatoes, are left over, the meat may be seasoned with a savory sauce, turned into a baking dish, the mashed potatoes slightly thinned with hot milk and then slightly thickened with flour, and used as a crust. This makes what we call a potato pie. Four tablespoonfuls of milk and four of flour would be a good allowance to each cupful of mashed potatoes.

# POTATOES — COLD BOILED

### Hashed Brown Potatoes

Chop two cold boiled potatoes rather fine, season with salt and pepper. Put a tablespoonful of butter in an ordinary sauté pan; when hot, put in the potatoes, smoothing and patting them down; stand over a moderate fire and allow them to cook undisturbed for at least eight minutes; then with a limber knife fold over one half as you would an omelet; stand again over the fire for about three minutes and turn at once on to a heated dish. These are exceedingly difficult to make. Directions must be carefully followed; the butter must be hot when you put in the potatoes; the whole must be packed firmly down so that it will not break when turning out.

### O'Brien Potatoes

Chop one green pepper rather fine. Chop sufficient red pepper to make two tablespoonfuls. Put two tablespoonfuls of butter in a frying pan, add the peppers, which must be sweet; shake until the peppers are soft, cover over four cold boiled potatoes, chopped rather fine, that have been seasoned with a teaspoonful of salt and a dash of pepper. Press them down as you do hashed brown potatoes, let them stand for a moment, stir them up, mix well, without breaking, and press down again. Let these stand until brown, fold over as you would an omelet and turn out on a heated platter.

### Potatoes au Gratin

To each four good-sized cold potatoes chopped fine allow a pint of cream sauce, to which you have added four tablespoonfuls of grated cheese; mix the potatoes with the sauce, turn them into a baking dish, dust with cheese, and brown in a quick oven.

Scalloped Potatoes

Cut cold boiled potatoes into dice; to each pint allow a half pint of cream sauce. Put a layer of the sauce in the bottom of a baking dish, put in the potatoes, season with salt and pepper, cover with another layer of cream sauce, dust the top with bread crumbs, dot here and there little bits of butter, and bake in a moderate oven until a golden brown.


Potatoes in Milk

Cold boiled potatoes may be cut into slices and cooked in milk in a double boiler until the whole is thoroughly heated; season with salt and pepper and serve.


Sweet Potatoes

Cold boiled or roasted sweet potatoes may be mashed while warm, seasoned with salt, pepper and butter and formed at once into croquettes; dip and fry the same as white potato croquettes.


Lyonnaise Potatoes

Cut cold boiled potatoes into small dice; to each pint allow a tablespoonful of butter; put the butter in an ordinary sauté pan, melt it, add a tablespoonful of chopped onion, shake until the onion is golden brown; throw in the potatoes, shake or toss over a hot fire until each piece is slightly browned; sprinkle lightly with a half teaspoonful of salt, a tablespoonful of parsley, and a dash of pepper; dish and serve.


Broiled Potatoes

Cut cold boiled potatoes into thin slices lengthwise; dip each slice in a little melted butter, dust it with salt and pepper, and broil it over a clear fire until a golden brown. For dyspeptics it is better to broil the potato first and add the butter after, as the heating of the butter renders it indigestible. Sweet potatoes may be broiled after this same rule, and would be less greasy than when fried.

### Vegetable Browned Hash

Chop two or three cold boiled potatoes rather fine, add an equal quantity of chopped carrot, and either string beans or peas, which ever you happen to have left over. You can add to this a cupful of stewed cabbage. Put two tablespoonfuls of butter into a shallow frying pan, mix the vegetables, put them into the butter, let them stand over a slow fire until they are browned thoroughly and crusted in the bottom. Fold one half carefully over the other, and press the two halves together; cook just a moment longer, and turn out on to a heated platter. This is a nice dish to serve with omelet and tomato sauce for luncheon or supper.

# CHEESE

The shells of Edam, or pine-apple cheese, after all the available cheese has been scooped out, will be used as a baking dish for stewed spaghetti or macaroni or rice. If care is taken, one shell may be used for three or four bakings. Boil the macaroni in plain water until tender; then drain, cut it into small pieces and add it to cream sauce. Pour this into the cheese shell, stand the shell on a piece of oiled paper in a baking pan and run into a moderate oven for fifteen or twenty minutes. Lift the shell carefully, put it on to a heated dish, and send at once to the table. After the macaroni has been taken out, the shell will be cleaned and put aside in a cold place for the next baking. There is just enough cheese imparted by the toasting of this shell to give ah agreeable flavor to the macaroni. Plain boiled rice may be heaped into the shells and steamed, or baked in the oven for a few moments.

Any scraps or bits of common cheese, when too hard and dry to serve on the table should be grated, put into a jar and put aside for cheese balls to serve with lettuce, cheese soufflé, for baked macaroni, or spaghetti, or for croquettes, cheese sauce, or Duchess soup.

### Cheese Soufflé

Put one cup of stale bread crumbs with a gill of milk over the fire for just a moment; take from the fire, add the yolks of three eggs, six tablespoonfuls of grated cheese, a half teaspoonful of salt and a dash of red pepper; stir in the well-beaten whites of the eggs; put into individual baking dishes; bake in a quick oven about eight minutes and send at once to the table.

### Cheese Balls

Grate or chop sufficient common cheese to make a half pint; add to it one pint of stale bread crumbs, a half teaspoonful of salt, a dash of red pepper and the whites of two eggs slightly beaten. Form these into small balls the size of an English walnut; dip in egg and then in bread crumbs and fry in smoking hot fat. These may also be made into small cylinder-shaped croquettes, and served with cream sauce.

### Duchess Soup

Put two tablespoonfuls of butter and a sliced onion in a saucepan; cook until the onion is soft and yellow; add to this two tablespoonfuls of flour, mix, and then add one quart of milk, a level teaspoonful of salt and a palatable seasoning of red pepper. Add six tablespoonfuls of grated cheese; stir in a double boiler until it is smoking hot; press through a fine sieve; reheat and send at once to the table.

### Cheese Pudding

Toast slices of stale bread until a golden brown and crisp to the center. This is best done in the oven. Put a layer of this toasted bread in the bottom of a baking dish; put over a quarter of a cup of grated or chopped cheese, sprinkle with salt and red pepper; then another layer of bread, another of cheese and the last of bread. Pour over sufficient milk to moisten the bread; bake in a quick oven fifteen minutes, and serve at once.

# SAUCES

All meat sauces are made after the same rule, changing the liquids to give varieties; for instance, one tablespoonful of butter (which means an ounce), and one tablespoonful of flour (a half ounce) are always allowed to each half pint of liquid. The butter and flour are rubbed together (better without heating), then the liquid added, cold or warm, the whole stirred over the fire until boiling. A half teaspoonful of salt and an eighth of a teaspoonful of pepper is the proper amount of seasoning.

### White Sauce

If you wish to make a white sauce, use one tablespoonful of butter, one tablespoonful of flour and a half pint of milk. Called also milk or cream sauce.

### Tomato Sauce

Tomato sauce will have the same proportions of butter and flour and a half pint of strained tomatoes.

### Sauce Bechamel

For sauce Bechamel, fill the cup half full of stock, then the remaining half with milk, giving again the half pint of liquid and usual quantity of butter and flour.

### Sauce Supréme

This is one of the nicest of all sauces to use with warmed-over chicken, duck or turkey. Rub together a tablespoonful of butter and one of flour, then add gradually a half pint of chicken stock; stir constantly until boiling, take from the fire, add the yolks of two eggs, strain through a fine sieve, add the seasoning, and serve immediately.

Sauces containing the yolks of uncooked eggs cannot be reboiled after the eggs are added.

### English Drawn Butter

For English drawn butter, use a tablespoonful of butter, a tablespoonful of flour, and a half pint of water. We usually have the water boiling, and add it gradually to the butter and flour, stirring rapidly. As soon as it reaches boiling point, take from the fire and add carefully another tablespoonful of butter. This may be converted into a plain

### Sauce Hollandaise

by adding with the last tablespoonful of butter, the yolks of two eggs, the juice of half a lemon, a teaspoonful of onion juice and a tablespoonful of chopped parsley.

### Brown Sauce

This is made by rubbing butter and flour together in the above proportions, then adding a half pint of stock; stir until boiling, add a teaspoonful of browning or kitchen bouquet and the usual seasoning of salt and pepper. To change the character of this sauce add garlic, onion, Worcestershire sauce, mushroom catsup, etc.

### Brown Tomato Sauce

An exceedingly nice sauce for Hamburg steaks. After you have taken the steaks from the pan, add a tablespoonful of butter and one of flour; mix. Fill your measuring cup half full of strained tomatoes, the remaining half with stock, making a half pint; add this to the butter and flour, stir until boiling, add a seasoning of salt and pepper and pour over the steaks.

## Roasted Beef Gravy

Roasted beef gravy, which really should be a sauce, is improved by adding a little tomato to the stock before adding it to the fat and flour. In roasting meats, we do not use butter for the sauce; there is always sufficient fat in the bottom of the pan. Pour from the pan all but one or two tablespoonfuls of fat (the amount required) and add to that the flour. A rounding tablespoonful of butter to which we refer weighs an ounce; of liquid fat, as in the pan, you must allow two even tablespoonfuls to the ounce; so, if you are going to make a half pint of sauce take out all but two tablespoonfuls of fat; add one tablespoonful of flour and then the half pint of water or stock.

## Browning

Plain burned sugar (caramel) may be used to color soups and sauces, thus saving the trouble of browning the flour or butter. It is also used as a flavoring for sweets. Put one cup of sugar, dry, into an iron saucepan. Stand it over a hot fire, and stir continually until it is reduced to a dark brown liquid. When it begins to burn and smoke, add hastily a cup of boiling water, stir and cook until a thin syrup-like mixture is formed. It must not be too thick. Bottle, and it is ready for use, and will keep any length of time.

## Kitchen Bouquet

Add one chopped onion and a teaspoonful of celery seed to one cup of dry sugar, and then proceed as for ordinary browning. Strain and bottle. A very good mixture under this name can be purchased at the grocers.

## Mushroom Sauce

Where just a few mushrooms are left over, either fresh or canned, they may be chopped fine and added to a brown sauce and served with steak or beef; or they may be chopped fine and added to a cream sauce and served with chicken or sweetbreads.

## Cold Meat Sauces

It is the fashion when one is serving cold meat to pass with it some condiment like Worcestershire sauce, mushroom, walnut or tomato catsup. Of course, these used in any great quantity are more or less injurious. A number of little left-overs in the house may be used to take their place, adding zest to the meat, and are more economical and more wholesome.

## Chopped Tomato Sauce

Peel a good-sized tomato, cut it into halves and press out the seeds; chop the flesh of the tomato fine, add a quarter of a teaspoonful of salt, a dash of pepper, or, if you have it, a little sweet pepper chopped fine; you may add also a little celery chopped very fine, or celery seed, and a teaspoonful of onion juice; rub your spoon with a clove of garlic, and mix the ingredients thoroughly; add a teaspoonful of lemon juice and dish. Pass and use as ordinary catsup.

## Grated Cucumber Sauce

Grate three or four large cucumbers; drain them on a sieve; to this drained pulp add a half teaspoonful of salt, a dash of red pepper, a teaspoonful of onion juice, a tablespoonful of lemon juice, and their stir in carefully two or three tablespoonfuls of very thick cream; if you can whip the cream a little first, so much the better. Cream may also be added to the tomato.

## Chopped Celery Sauce

Chop fine sufficient celery to make a half pint; season it with a quarter of a teaspoonful of salt, a teaspoonful of onion juice, a dash of pepper. Rub the spoon with garlic, mix thoroughly, stir into it the yolk of an egg that has been beaten light with two tablespoonfuls of cream; add a few drops of lemon juice or tarragon vinegar and serve.

Cream Horseradish Sauce

This is one of the most delightful sauces to serve with left-over meats, especially beef. Press from the vinegar four tablespoonfuls of horseradish, add a quarter of a teaspoonful of salt, and work in the yolk of an egg. Whip six tablespoonfuls of cream to a stiff froth, stir it gradually into the horseradish and dish at once.


Pudding Sauces

The simple method of making a pudding sauce is to add to a half cup of sugar, a tablespoonful of flour; mix thoroughly, and then add hastily a half pint of boiling water; boil for a moment and pour while hot into one well-beaten egg, beating all the while. This may now be seasoned with any flavoring, as orange, lemon or vanilla.

To change the character of this sauce, a tablespoonful of butter may be added. Where butter enters largely into the composition of a pudding sauce, it is better that it should be beaten to a cream, the sugar added gradually, then the egg and last the liquor. Heat it over a double boiler just at serving time, or the froth will float on the surface and the liquid be rather dense at the bottom.

Melted sugar with lemon juice and a little water is called sugar sauce.

# SALADS

There comes a time during the week, even in careful housekeeping, when there is an accumulation of little things, a few olives, a slice or two of beet, perhaps two or three pieces of cooked carrot, a cold potato, a tiny little bit of cold fish, or cold meats, and not more than a tablespoonful or two of aspic jelly; these may all be utilized in a

Russian Salad

Chop or cut carefully the vegetables; mix together, add two or three tablespoonfuls of toasted piñon nuts, and the meat and fish; dish on lettuce leaves, or, if you have tomatoes, peel and take out the centers, and fill the salad into the tomatoes. Serve with French or mayonnaise dressing; garnish with blocks of aspic jelly.

# CEREALS

Cold boiled rice left over may be mixed with a small quantity of meat, and used for stuffing tomatoes or egg plant; or it may be reheated or made into pudding, or added to the muffins for lunch, or added to the corn bread.

A cup of oat meal or cracked wheat or wheatlet may also be added to the muffins or ordinary yeast or corn breads. These little additions increase the food value, make the mixture lighter, and save waste.

### Southern Rice Bread

Separate two eggs, beat the yolks until light, and add one cup (a half pint) of milk; add a tablespoonful of melted butter, a half teaspoonful of salt, and one and a half cups of corn meal; beat thoroughly, and stir in one cup of cold boiled rice; add a teaspoonful of baking powder; beat for two or three minutes; stir in the well-beaten whites of the eggs, and bake in a thin sheet in an ordinary baking pan.

### Rice Muffins

Separate two eggs; add to the yolks one cup of milk and a cup and a half of white flour; beat thoroughly, add a half teaspoonful of salt, a teaspoonful of baking powder and one cup of cold boiled rice; stir in the well-beaten whites, and bake in gem pans in a quick oven twenty minutes.

### Rice Croquettes

To make cold boiled rice into croquettes, the rice must be reheated in a double boiler with a gill of milk and the yolk of an egg to each cup; you may season with sugar and lemon or salt and pepper, and serve as a vegetable. Form into cylinder-shaped croquettes; dip in egg and bread crumbs, and fry in smoking hot fat.

### Simple Rice Pudding

Put into a double boiler one quart of milk; allow it to cook for thirty minutes; then add two tablespoonfuls of sugar, a grating of nutmeg, and one cup of cold boiled rice; turn this into a baking pan, and bake in a quick oven thirty minutes. Serve cold. Raisins may be added when it is put into the baking pan.

### Lemon Rice

Into one cup of cold boiled rice stir one pint of milk; beat the yolks of three eggs with a half cup of sugar together until light; add to them the rice and milk; add the grated yellow rind and the juice of one lemon. Turn this into a baking pan; bake in a moderately quick oven twenty to thirty minutes. Beat the whites of the eggs to a stiff froth, add three tablespoonfuls of powdered sugar, and beat again. Heap these over the pudding, dust thickly with powdered sugar; return to the oven to slowly brown; serve cold.

### Paradise Pudding

Pare, core and grate three apples. Separate three eggs; add to the yolks four tablespoonfuls of sugar; beat until light; add a grating of nutmeg and a teaspoonful of lemon juice; stir in a half cup of cold boiled rice; mix with this quickly the apples, and beat well; add a half cup of milk; turn into a baking dish, and bake for thirty minutes. Make a meringue as in preceding recipe, from the whites of the eggs; heap it over the top, and brown. This pudding may be served warm or cold.

### Compote of Pineapple

Throw a pint of boiling water over one cup of cold boiled rice; stir for a moment; drain, and stand at the oven door. Have ready, picked apart, one small pineapple; add to it a half cup of sugar; heat quickly, stirring constantly. Arrange the rice in the center of a round dish, making it into a mound, flat on top; heap the pineapple neatly on this; pour over the syrup, and send at once to the table. Small quantities or different kinds of fruits that have been left over may be blended and used in this way.

### Monday Pudding

Cut bits of whole wheat bread into dice. Use a half cup of any fruit that may have been left over, prunes, raisins, chopped dates or candied fruit. Grease an ordinary melon mold; put a layer of the bread in the bottom, then a layer of the fruit, and so continue until you have the mold filled. Beat three eggs, without separating, with four tablespoonfuls of sugar; add a pint of milk; pour this carefully over the bread; let it stand for ten minutes; then put the lid on the mold, and steam or boil continuously for one hour. Serve with lemon or orange sauce.

### Apple Farina Pudding

Pour the left-over breakfast porridge into a square mold and stand it aside. At luncheon or dinner time cut this into thin slices, cover the bottom of a baking dish with these slices, and cover these with sliced apples, and so continue until you have the ingredients used, having the last layer apples. Beat an egg, without separating, until light, add a half cupful of milk and a saltspoonful of salt, then stir in a half cupful of flour. When smooth pour this over the apples and bake in a quick oven a half hour. Serve with milk or with hard sauce.

### Cranberry Farina Pudding

2 cupfuls of cold left-over farina porridge 1/2 cupful of cranberries 1/2 cupful of sugar

It is wise to pour the porridge into a mold as soon as you finish breakfast. At serving time turn this out in a glass dish, pour over the cranberry that has been pressed through a sieve; dust thickly with the sugar. Stir the remaining sugar into a half pint of milk or cream and serve as a sauce with the pudding.

### Plain Farina Pudding

2 cupfuls of milk 1/2 cupful of sugar 2 eggs 1 cupful of left-over farina or cream of wheat 1 teaspoonful of vanilla

Put the milk in a double boiler, add the sugar and cold farina porridge. Stir until thoroughly hot, then add the eggs, well beaten, and the vanilla. Turn into a baking dish and run in the oven until brown. Serve cold, with milk or cream.

### Farina Gems

2 eggs 1 cupful of milk 1 cupful of cold boiled farina 1 cupful of flour 4 level teaspoonfuls of baking powder 1/2 teaspoonful of salt

Separate the eggs, add the milk and stir this gradually into the cold farina. When smooth add the salt, baking powder and flour, mixed. Beat, and then fold in the well-beaten whites of eggs. Bake in gem pans in a quick oven a half hour.

### Hominy Pone

1 cupful of boiled hominy 1 cupful of white corn meal 2 cupfuls of milk 2 level tablespoonfuls of butter 2 eggs 1/2 teaspoonful of salt

If the hominy is cold left-over hominy, add to it the milk, and when thoroughly smooth add the eggs, well beaten, then the butter, melted, and the corn meal. Pour into a greased pan and bake in a very hot oven about twenty to twenty-five minutes.

### Oat Meal Muffins

The ordinary muffin recipes, which are always about alike, no matter what flour is used, may have added to them a cup of well-cooked oat meal; for instance, separate two eggs as for rice muffins; add to the yolks a cup of milk; then add one and a half cups of whole wheat flour; beat thoroughly; add a teaspoonful of baking powder; beat again; add one cup of well-cooked oat meal, or you may substitute wheatlet or any of the breakfast cereals; fold in the whites of the eggs, and bake in gem pans in a quick oven twenty to thirty minutes.

Sandwiches

Little bits of fruit, crisp pieces of celery, cold meats of all kinds, may be chopped, properly seasoned, and used for making fruit, vegetable and meat sandwiches.

# VEGETABLES

String beans, cauliflower, carrots, beets, peas and even a cold boiled potato may all be cut into neat pieces, mixed together, and served on lettuce leaves, dressed with French dressing as a salad. One cold boiled beet may be used as a garnish for a potato salad. String beans, if you have sufficient quantity, may be served alone as a salad.

Stuffed Egg Plant

Throw a good-sized egg plant into a kettle of boiling water; boil ten minutes; when cold cut into halves and with a blunt knife scoop out the center. Chop this scooped-out portion fine, mix with it an equal quantity of finely-chopped uncooked meat, add a grated onion, a clove of garlic mashed, a teaspoonful of salt, a little chopped parsley, if you have it, and a dash of pepper. Fill this into the egg plant shells, stand them in a baking pan, add a cup of stock and a tablespoonful of butter, bake slowly one hour, basting every ten minutes.

Cucumbers

Raw cucumbers are easily wilted, and are then unfit for serving. Soak them in pure cold, unsalted water until serving time. Pass French dressing in a separate dish. In this way the "left-overs" may be placed in the refrigerator and used next day as an addition to the dinner salad.

Left-Over Tomatoes

A half cup of stewed tomatoes may be used with stock for brown tomato sauce, or for making a small dish of scalloped tomatoes, helping out at lunch when perhaps the family is less in number. The

Italians boil down this half cup of tomatoes until it has the consistency of dough; then press through a sieve, add a little salt, pack down into a jelly tumbler and stand in the refrigerator to use as flavoring. A tablespoonful in a soup, or in an ordinary sauce, or mixed with the water for baked beans, or added to the stock sauce for spaghetti or macaroni, adds greatly to the flavor as well as appearance.

### Corn Oysters

6 ears of cold boiled corn 2 eggs 1 cupful of milk 1/2 cupful of flour 1/2 teaspoonful of salt 1 saltspoonful of pepper

Score the corn, press it out, add the eggs, well beaten, and the oil or butter; then stir in the milk, salt and pepper. Sift the flour, stir it in, and drop by spoonfuls into shallow hot fat.

### Chicken Corn Pie

6 ears of cold cooked corn 4 eggs 1 level tablespoonful of butter, melted 1 cupful of milk 1 teaspoonful of salt 1 saltspoonful of pepper 1 young chicken

Score the corn and with a dull knife press it out. Carefully beat the eggs, without separating, until light, add the milk, melted butter, salt and pepper. Pour this into a casserole mold or pudding dish. Have the chicken drawn and disjointed; make two pieces of the breast, cut it into four pieces, dust with salt and pepper, brush with melted butter. Lay the chicken on top of this mixture and stand the baking dish in a moderately quick oven about one hour. Serve in the dish in which it was cooked. Some prefer to broil the chicken on the bone side before they put it into the pudding, the pudding may be baked, and then put it in the pudding and brown it with the pudding. This is a good way to use cold left-over corn, and cold bits of chicken may be used in the place of the fresh chicken.

### Green Corn Cakes

4 ears of left-over cooked corn 1 egg 2 tablespoonfuls of milk 1 tablespoonful of melted butter 1/2 cupful of flour 1/2 teaspoonful of salt

Score the corn, press out the cooked pulp, add to it the beaten egg, milk, melted butter and salt. Stir in the flour, and drop by tablespoonfuls into a little thoroughly heated fat.

# FRUITS

Small quantities of fruit that are not sufficiently sightly to put again on the table may be put aside and made into fruit pot-pie. All sorts of fruits may be blended. Put them into a saucepan, and to each pint of this fruit allow one quart of water and a palatable seasoning of sugar, and you may flavor it with a little grated lemon or orange rind; bring to boiling point. During this time put one pint of flour into a bowl, add a half teaspoonful of salt and a teaspoonful of baking powder. Beat one egg until light, add to it a half cup of milk, then add this to the flour; there should be just enough to moisten and make a dough. Take this out on the board, knead lightly, roll out and cut into biscuits. Put these biscuits over the top of the fruit; cover the kettle and cook slowly for fifteen minutes; do not lift the lid during the cooking. Serve hot with plain milk or cream, or with a hard sauce made from sugar and butter.

Fruit Soufflé

Beat the whites of six eggs until light, but not dry; add three tablespoonfuls of powdered sugar; mix quickly; line the bottom of the baking dish with any sort of fruit, such as chopped dates or figs, or left-over candied fruits or preserves. Heap over the whites of the eggs, dust thickly with powdered sugar, and bake in a hot oven for five minutes. Serve immediately. To give variety, where stale biscuits or bread, or sponge cake are left over, line the bottom of the dish with the stale bits; pour over enough milk to moisten, put in a layer of fruit and the whites of the eggs as above.

Fruit Jambolaya

Put one cupful of cold boiled rice in a little sieve or colander and stand it over the tea kettle where the steam will pass through it. Chop fine any left-over fruit at hand, an apple, pear, plum, banana, and the pulp of an orange; they may be all mixed together and slightly sweetened. Put a little of the rice into four serving dishes, put in the center of each a tablespoonful of the chopped fruit and send to the table. This is rather nice for children, and is a good way to use up both the rice and the fruit, as it makes a good combination.

### Plain White Cake

Beat a quarter of a cup of butter to a cream; add gradually one and a half cups of sugar. Sift two cups of flour with a teaspoonful of baking powder; measure a half pint of water; add a little water and a little flour, and so continue until the ingredients are used; beat thoroughly, then stir in the well-beaten whites of five eggs. Bake in a loaf or layers. Put layers together with chopped fruit, soft custard, or a soft icing.

### Chicken Muffin Cases

Boil together a half pint of water and two tablespoonfuls of butter, add hastily a half pint of sifted flour, stir over fire until a smooth dough is formed. Take from the fire and when cool, add one unbeaten whole egg; beat, add another and so continue until four eggs have been added. Bake in gem pans until light and hollow, about a half hour. This quantity will make twelve. Cut a round from the top and fill the muffin with any creamed mixture.

### To Make Cocoanut Milk

Cover one quart of grated cocoanut with one pint of boiling water. Stir and mash; strain and press. The milk thus produced may be used for curries. Throw away the pulp.

# SOUR MILK AND CREAM

Corn Cake

2 eggs 1 cupful of thick sour milk 1 level teaspoonful of baking soda 2 cupfuls of corn meal 3/4 cupful of white flour 2 cupfuls of sweet milk 3 level teaspoonfuls of baking powder

Beat the eggs until very light, without separating. Moisten the soda in two tablespoonfuls of cold water, stir it into the cupful of sour milk; add this to the eggs, then add the meal and beat thoroughly. Sift the baking powder and flour; stir these into the other mixture, and then add the two cupfuls of sweet milk. Pour into a shallow greased pan and bake in a moderately quick oven about three-quarters of an hour. This should have a custard on top.


Sponge Corn Cake

1 cupful of corn meal 1/2 cupful of flour 1 cupful of thick sour milk 2 eggs 1 level tablespoonful of butter, melted 1/2 teaspoonful of salt 1/2 teaspoonful of baking soda

Moisten the soda in a tablespoonful of water and stir into the thick sour milk. Separate the eggs; beat the yolks, add the sour milk, with the butter, melted, corn meal and flour. Beat thoroughly, then fold in the well-beaten whites, add salt and bake in a shallow greased pan in a quick oven a half hour.


Old Virginia Batter Cakes

2 eggs 1 cupful of sour milk 1 cupful of water 2 cupfuls of white corn meal 1 cupful of flour 1/2 teaspoonful of salt 1 level teaspoonful of baking soda 1 teaspoonful of baking powder

Beat the eggs, without separating, until very, very light. Dissolve the soda in a little water, add it to the sour milk; stir until this is well mixed, add it to the egg; add the water, the corn meal, salt and flour sifted with the baking powder. Mix thoroughly and bake on a very lightly greased griddle.

Plain Corn Dodgers

1 egg 1/2 teaspoonful of salt 1 cupful of thick sour milk 1 level teaspoonful of baking soda 1 cupful of corn meal 1/2 cupful of flour

Beat the egg, without separating. Dissolve the soda and add it to the sour milk; add this to the egg; add the salt, then the corn meal and flour. Beat until well mixed, and drop by spoonfuls in a shallow pan in which you have a little bacon or ham fat. When cooked on one side, turn quickly and cook on the other.

# INDEX

Anchovy, Mutton with
Apple Farina Pudding
    Snow

Baked Sardines
Balls, Cheese
    Curry
    English Chicken
Barbecue of Cold Beef
Batter Cakes, Old Virginia
Beauregard Eggs
Bechamel Sauce
Beef, Cold, Barbecue of
Beef — Cooked
  Barbecue of Cold
  Bresleau
  Croquettes
  Fritters
  Gobbits
  Minced on Toast
  Panada
  Potato Dumplings
  Ragout
  Rechauffee
  Salt Hash No. 1
      No. 2
  Steak Pudding
Beef Croquettes
  Fritters
  Gravy, Roasted
  on Toast, Minced
  Panada of

Rechauffee of
　　Salt Hash No. 1
　　　　No. 2
　　Steak Pudding
　　Timbale
Beef — Uncooked
　　Brown Stew
　　Cannelon
　　Hamburg Steaks
　　Kibbee
　　Timbale
Bobotee
Bordelaise Duck
Boudins
Bouquet, Kitchen
Bread
　　　and Butter Custard
　　　Croquettes
　　　Muffins
　　　Southern Rice
Bresleau
Broiled Potatoes
Browned Hash, Vegetable
Browning
Brown Sauce
　　Stew
　　Tomato Sauce
Butter, English Drawn


Cake, Corn
　　Gold
　　Plain White
　　Sponge Corn
Cakes, Green Corn
　　　Old Virginia Batter
Canapés
Cannelon
Cases, Chicken Muffin

Casserole
Celery Sauce, Chopped
Cereals
Cheese
    Balls
    Pudding
    Soufflé
Chicken Balls, English
Chicken — Cooked
  Casserole
  Creamed Hash on Toast
  Cutlets
  Indian Hash
  Mock Terrapin
  Supréme
Chicken Corn Pie
    Cutlets
    Legs, Deviled
    Muffin Cases
    Supréme
Chicken — Uncooked,
  Deviled Legs,
  English Balls,
  Timbale,
Chopped Celery Sauce,
    Tomato Sauce,
Cocoanut Milk, To Make,
Cold Beef, Barbecue of,
    Boiled Potatoes,
    Meat Sauces,
Compote of Pineapple,
Cooked Beef,
    Chicken,
    Fish,
    Mutton,
Corn Cake,
    Sponge,
  Cakes, Green,
  Dodgers, Plain,

Oysters,
   Pie, Chicken,
Cranberry Farina Pudding,
Cream Horseradish Sauce,
Creamed Hash on Toast,
Croquettes, Beef,
        Bread,
        Egg,
        Fish,
        Potato,
        Rice,
Cucumber Sauce, Grated,
Cucumbers,
Curry Balls,
    of Mutton,
Custard, Bread and Butter,
Custards, Potato,
Cutlets, Chicken,


Deviled Chicken Legs,
Dodgers, Plain Corn,
Drawn Butter, English,
Duchess Soup,
Duck Bordelaise,
Dumplings, Potato,


Egg Croquettes,
   Plant, Stuffed,
Eggs,
    Beauregard
    Whites of,
English Chicken Balls,
     Drawn Butter,


Farina Gems,
     Pudding, Apple,
           Cranberry,

Plain,
Fish à la Crême,
Fish — Cooked,
  à la Crême,
  Baked Sardines,
  Canapés,
  Croquettes,
French Lamb Stew,
Fritters, Beef,
Fruit Jambolaya,
    Soufflé,
Fruits,


Game,
Garnishing, Potato Roses for,
Gems, Farina,
German Slaw,
Gobbits,
Gold Cake,
Grated Cucumber Sauce,
Gravy, Roasted Beef,
Green Corn Cakes,


Hamburg Steaks,
Hash, Creamed, on Toast,
    Indian,
    Salt Beef No. 1,
        No. 2,
    Vegetable Browned,
Hashed Brown Potatoes,
Hollandaise Sauce,
Hominy Pone,
Horseradish Sauce, Cream,


  Indian Hash,

  Jambolaya, Fruit,

Kibbee,
Kitchen Bouquet,
Klopps,

Lamb Stew, French
   with Tomatoes
Left-Over Tomatoes
Lemon Rice
Little Puddings à la Grand Belle
Lyonnaise Potatoes

Meat
   Sauces, Cold
Milk, Cocoanut, To Make
   Potatoes in
Minced Beef on Toast
Mock Terrapin or à la Newburg
Monday Pudding
Muffin Cases, Chicken
Muffins, Bread
     Oat Meal
     Rice
Mushroom Sauce
Mutton — Cooked
   Bobotee
   Boudins
   Curry of
   French Stew
   Klopps
   Pilau
   Salad
   Stew with Tomatoes
   with Anchovy
Mutton, Curry of
   Salad
Mutton — Uncooked
   Curry Balls

Mutton with Anchovy

Oat Meal Muffins
O'Brien Potatoes
Old Virginia Batter Cakes
Oysters, Corn

Panada of Beef
Paradise Pudding
Pie, Chicken Corn
Pilau
Pineapple, Compote of
Plain Corn Dodgers
    Farina Pudding
    White Cake
Pone, Hominy
Potato Croquettes
    Custards
    Dumplings
    Puff
    Roses, for Garnishing
Potatoes
    au Gratin
    Broiled
    —Cold Boiled
    Hashed Brown
    in Milk
    Lyonnaise
    O'Brien
    Scalloped
    Stuffed
    Sweet
Pudding, Apple Farina
    Beef Steak
    Cheese
    Cranberry Farina
    Monday
    Paradise

    Plain Farina  
    Sauces  
    Simple Rice  
    Steak  
Puddings, Little à la Grand Belle  
Puff, Potato  

Ragout  
Rechauffee of Beef  
Rice Bread, Southern  
    Croquettes  
    Lemon  
    Muffins  
    Pudding, Simple  
Roasted Beef Gravy  
Roses, Potato, for Garnishing  
Russian Salad  

Salad, Mutton  
    Russian  
Salads  
Salt Beef Hash, No. 1  
        No. 2  
Sandwiches  
Sardines, Baked  
Sauce, Bechamel  
    Brown  
    Tomato  
    Chopped Celery  
    Tomato  
    Cream Horseradish,  
    Grated Cucumber  
    Hollandaise  
    Mushroom  
    Supréme  
    Tomato  
    White  
Sauces,

Cold Meat
Pudding
Scalloped Potatoes
Simple Rice Pudding
Slaw, German
Snow, Apple
Soufflé, Cheese
    Fruit
Soup, Duchess
Sour Milk and Cream
   Corn Cake
   Old Virginia Batter Cakes
   Plain Corn Dodgers
   Sponge Corn Cake
Southern Rice Bread
Sponge Corn Cake
Steak Pudding,
   Beef
Steaks, Hamburg
Stew, Brown
   French Lamb
   Lamb, with Tomatoes
Stock
Stuffed Egg Plant
    Potatoes
Supréme Chicken
    Sauce
Sweet Potatoes


Terrapin, Mock
Timbale
    Beef
To Make Cocoanut Milk
Tomato Sauce,
   Brown
    Chopped
Tomatoes, Lamb Stew with

Left-Over

Uncooked Beef
    Chicken
    Mutton

Vegetable Browned Hash
Vegetables

White Cake, Plain
   Sauce
Whites of Eggs

www.ingramcontent.com/pod-product-compliance
Lightning Source LLC
Chambersburg PA
CBHW070314230526
45470CB00002B/865